A Guide to

EXECUTIVE
PROTECTION

Nick Nicholson

Perpetuity Press Ltd
PO Box 376 Leicester LE2 1UP UK
Telephone: +44 (0) 116 221 7778
Fax: +44 (0) 116 221 7171

214. N. Houston
Comanche Texas 76442 USA
Telephone: 915 356 7048
Fax: 915 356 3093

Email: info@perpetuitypress.com
Website: http://www.perpetuitypress.com

First Published 2001

British Library Cataloguing in Publication Data.
A catalogue record for this book is available from the British
Library.

A Guide to Executive Protection
by Nick Nicholson

ISBN 1 899287 59 0

About this Guide

This guide is aimed at Executive Services Managers who want and need to understand the requirements of Executive Protection.

The contents of the guide are based on a variety of information collected from various professional groups who, as their primary responsibility, provide protection to people. The information used and outlined in the guide is the result of a research study and data collection by the author over the last five years. The primary focus of the study was to conduct a needs assessment of the type of knowledge and skills and abilities an Executive Protection Specialist should possess. The groups identified for this study included US Federal Agencies having a protection team, and private security organizations that conduct executive protection teams as a primary function of their security responsibilities. No classified information was used in the study and all attempts to identify agency specific techniques have been eliminated. Once the needs assessment was completed the author began to systematically categorize the type of tasks a protection specialist should be able to perform. Once the task list was developed, professional protection specialists[1] were surveyed and asked to identify how often (frequency) a task was performed, how important (critical) the task was and the difficulty level of the task.

The completed surveys were used to prioritize the tasks to be performed and the type of knowledge and skill level required to be successful. As an example, one task identified on the survey was the ability to drive. The survey results indicated that professionals believed the frequency level of this task to be high and the level of importance to be high, which makes it a critical task to be trained, and a requirement for an executive protection specialist.

1. Professional is defined as an individual who has at least five continuous years of service in the field, not necessarily with the same principal, and who attended some degree of training in executive protection.

Therefore, driving would be listed as an immediate requirement for an EPS, and should be a skill area requiring ongoing refresher and advance training. Other skill areas identified included interpersonal skills, report writing, intelligence gathering, and emergency medical treatment, to name a few.

The information presented in the guide represents the type of knowledge and skill areas required to provide protection services. Each section illustrates the task to be performed and provides the reader with a general checklist of the type of information or skill required. To help remember the different task areas the acronym 'P.R.O.T.E.C.T.' has been created. Each task is represented by the letters P-Plans, R-Response, O-Organization, T-Technical, E-Emergency, C-Communications, and T-Threat Assessment, but is not necessarily arranged in priority.

Since the purpose of this guide is to provide an overview of the knowledge and skill levels required to be an EPS, each section does not contain a complete list of characteristics. The size of a protective team will vary depending on the threat and the client's desires. Large government teams are not typically found in private sector organizations.

Contents

Section 1
Plans

INTRODUCTION

One of the most effective means to ensure that the Executive Protection Specialist (EPS) performs the tasks as required is to establish effective plans and ensure that personnel use them. The only way to be prepared for the unexpected is to plan for it. Although considered the most important, it is usually the least glamorous part of the profession. It takes time to gather information through interviews and observations as well as training on what items should be noted in a useful plan. Figure 1 depicts the types of plans that should be in place for the organization.

OFFICE SECURITY PLAN

This plan is designed to identify the environment in which the principal works in and the type of security that is required and should by normal course of business be maintained consistently. What plans are in place to ensure that temporary people are briefed on the security activities and how to respond in an emergency?

The Office Security Plan (OSP) begins with a total security assessment of the building housing the principal. The survey plan should contain the address of the building and a physical description of the location, such as intersecting streets and roadways. Appendix A illustrates structural items that should be considered in the OSP.

In addition to general information about the building, the OSP should contain the following security related information.

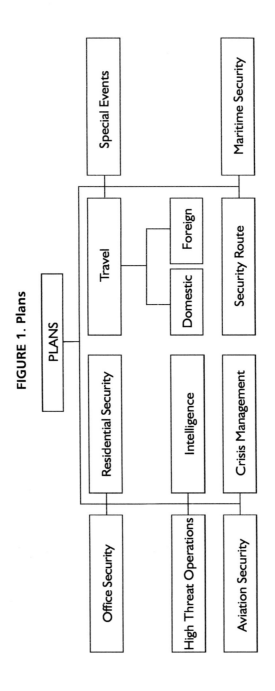

FIGURE 1. Plans

The type of access or identification credentials required for:

- personnel;

- vehicles;

- visitors;

- principals;

- egress/regress areas to include garages.

Security response contact data should be recorded, and should include the number of personnel on duty, their weapons capabilities and the estimated timescale required to respond to specific locations within the building/office area. The OSP should contain floor plans of the building, especially the layout of the principal's common areas. Security technologies such as sensors, video cameras and duress switches should be annotated on the floor plan.

In the principal's immediate office areas, access control procedures should be outlined in the OSP as well as guidelines for responding to emergency situations. *Unusual Incident* checklists should be kept close to the receptionist area and reviewed everyday to maintain familiarization. As in any plan, the staff should be trained on the plan and exercise the plan at least every six months or when major changes are made.

RESIDENTIAL SECURITY PLAN

The Residential Security Plan (RSP) is similar to that of the OSP, but the depth of the plan is expanded to cover not only the principal but also his/her family. Since many principals typically own numerous properties, the RSP may vary depending on the function the property serves the principal. Appendix B depicts the structural items that should be considered in the RSP.

Residential staff members should be screened and extensive background examinations conducted. Staff members are often left unchecked for long periods of time and can develop habits contrary to good business practices. Staff members responsible for purchasing merchandise for residents have been known to work 'special deals' with vendors and suppliers. Internal audits should be conducted periodically to maintain appropriate oversight of staff members.

The staff should be trained in emergency procedures to include first responder (first aid), fire suppression and emergency evacuation. Complete safety checklists addressing these issues should be posted throughout the residence. The staff should conduct emergency drills or exercises at least twice a year and always prior to the family moving into the residence. Unannounced exercises should be conducted by the supervising EPS. Access control procedures are outlined in the RSP as well as guidelines for responding to emergency situations.

The RSP will cover the residence and all associated outbuildings. The survey plan should contain the address of the residence and a physical description of the location, such as intersecting streets and roadways.

Additionally, the RSP should contain the following security related information.

The type of access or identification credentials required for:

- staff;

- vehicles allowed on the grounds;

- guests and visitors;

- family members;

- egress/regress from each building structure and the grounds.

Security response contact data should be recorded, and should include the number of personnel on duty, their weapons capabilities and the estimated timeline required to respond to specific locations on the grounds. The RSP should contain floor plans of the buildings, especially the layout of the principal's living areas. Security technologies such as sensors, video cameras and duress switches should be annotated on the floor plan.

HIGH THREAT OPERATIONS PLAN

The High Threat Operations Plan[2] (HTOP) has one goal, to provide an immediate, planned response to an imminent threat. Elements of the HTOP are selected from each of the other plans and survey information prepared before each principal movement. Appendix C lists the HTOP elements that should be considered.

AVIATION SECURITY PLAN

The Aviation Security Plan (ASP) is developed for each trip the principal is scheduled to take. Airports used consistently by the principal should be surveyed first and relationships with the local law enforcement and airport personnel developed. Anonymity has its advantages, but the ability to gain cooperation through familiarization has its advantages when attempting to expedite the principal. Appendix D identifies the type of information contained in the ASP.

When selecting a private aircraft, the EPS manager should become familiar with the type of aircraft being used, the model year and the tail number. Flight maintenance schedules are based on the number of hours on the aircraft. The number of hours before the next regular maintenance should be identified to ensure no required maintenance is needed while the principal is using the aircraft.

2. High Threat Operations exist when the probability of an attack is higher than normal, based on collected intelligence.

The ASP should always contain departure/arrival information such as times, location and contact numbers. The EPS manager or representative should maintain contact information on the flight crew throughout the trip.

INTELLIGENCE PLAN

The Intelligence Plan (IP) outlines the means by which intelligence data is collected and information formulated into a usable document. The IP identifies the resources that are used, the type of information available from each type of source, and databases it for future retrieval. The topic areas covered by the IP include:

- list of external sources of information:

  - journals;
  - periodicals;
  - commercially available country threat reports;
  - government publications (ie, US Department of State);
  - internet;

- list of internal sources of information:

  - periodic debriefs of EPS team and staff;
  - intranet sites that identify corporate-wide activities;
  - security reports and related analysis information;
  - phone and mail contact database identifying potentially threatening individuals;
  - unwelcome visitor reports;

- networking:

  - contacts for local, state, and federal law enforcement agencies;
  - contacts of people performing similar tasks at other organizations. Do not disregard some contacts because they are competitors;

– contacts, internal or external, who give advice on medical and legal issues. As an example, who is experienced in the field of determining the behavior of a person writing threatening letters?

Regardless of the data collected, it is worthless unless the database is maintained continuously, the information is easily retrievable, and it is actually reported in a format that will be used. If none of this occurs, there is no need to waste time developing an IP or even gathering threat information as outlined in Section 7, Threat Assessment.

TRAVEL PLAN (DOMESTIC AND FOREIGN)

The development of the Travel Plan (TP) is a coordinated effort between the EPS manager and typically the Executive Services organization. As a minimum a travel plan should contain the following information:

- destination of each trip segment and telephone code numbers;

- hotel address and contact information (persons and numbers) and fax numbers;

- type of accommodation requested, confirmation number and name of several hotel managers (day/night shifts, catering, etc.) responsible;

- the name in which the reservations were made. This may change depending on the public profile of the principal;

- business center facilities available (fax, phone, copier, computer/internet terminal, availability after hours).

The Airport Security Plan can provide information such as the handling agent, their telephone/fax numbers, points of contact, customs and immigration facilities, and applicable restrictions (visa/passport and immunization requirements). Complete information on car services should be annotated in the plan and should include the name of the service, driver, contact data, the driver's license number, and 24/7 contact information. The principal may have a car preference, which should of course be considered in finding the right service. The number of passengers and amount of luggage will influence the type of automobile required.

The TP will contain a list of all businesses to be visited and related contact information as well as contact information for local restaurants and emergency services. Emergency service information includes:

• local police/fire contact numbers and location in relation, always, to the principal's location;

• hospital contact numbers, location in relation to the principal's location, and specialization, ie. trauma, acute care;

• doctor/dentist contact numbers, especially the names and information about physicians that can provide special needs, such as a pediatrician or veterinarian.

Foreign travel presents a special challenge to the EPS manager. As well as the type of information required for domestic travel, additional information gathered will include:

• medical facilities that can meet the needs of the principal;

• customs requirements and additional entrance requirements such as country visas;

• up-to-date information on any political situations and/or internal conflicts;

• identification of local national holidays that may pose a problem;

- special requirements for emergency evacuation for either medical purposes or potentially threatening situations;

- location of Embassy, especially important if notifying the local authorities would place the principal in greater danger.

SPECIAL EVENTS PLAN

As the EPS manager you will be asked to develop a Special Events Plan (SEP) for the principal. The types of events may range from a concert appearance to autograph signing sessions. Regardless of the type of special event to be coordinated there are some common aspects that should be considered. Appendix E illustrates the commonality in planning such events.

Other types of special events that the EPS manager will be called upon to plan and oversee include autograph signings, funerals, weddings, sporting events and family gatherings like holidays and birthdays, each presenting their own protection issues.

SECURITY ROUTE PLAN

The Security Route Plan (SRP) is developed to ensure that as much information as possible is available to the EPS team prior to transporting the principal. The EPS manager should travel common routes used routinely on a consistent basis to ensure that changes are identified and communicated to EPS team members. It doesn't impress the principal to be driven home from the airport and run into a construction zone (traffic at a dead stop) that was established several days earlier, when the EPS stationed at the residence should have known the situation and alerted the team. Appendix F outlines the type of information that should be included in the route plan.

It is critical to perform regular surveillance detection sweeps of common routes to ensure that an adversary isn't surveying the principal. During these sweeps the EPS

15

manager should look for anything out of the ordinary, such as people or vehicles in the area that they haven't seen before.

CRISIS MANAGEMENT PLANS (SEE EMERGENCIES)

Crisis Management Plans (CMP), developed for a principal, should be in line with the corporate policies that exist for the principal. If possible, the EPS manager should be a part of the Crisis Management Team. The team typically consists of senior management personnel who are able to act decisively under stress, each representing different disciplines within the corporation, such as human resources, legal, operations, finance and security. The functions performed by the team include:

- assessing the nature and level of threat to the corporate asset;

- determining the vulnerability associated with thecompany, personnel, and facilities;

- developing organizational structures to assist in mitigating the crisis or critical incident;

- evaluating the CMP on a regular basis, and ensuring that new members are integrated into the team;

- maintaining up-to-date CMPs.

Different types of plans are developed to cover a wide variety of potential incidents. Common situations that are addressed include:

- emergencies in the principal's office, residence, and at special events;

- hijacking of corporate aircraft;

- kidnapping and extortion;

- evacuation of personnel during an emergency;

- bomb threats, including the latest potential threat of chemical and biological weapons.

MARITIME SECURITY PLAN

The main purpose of a Maritime Security Plan (MSP) is to establish an environment in which the principal and his/her family, guests, staff and potential cargo can traverse by waterways safely and securely. Internationally, ports and sea craft are experiencing dramatic increases in various forms of maritime crime, including theft and in transit hijacking of containers and ships.[3] Additionally, sea robbery, stowaways (including alien smuggling) and drug smuggling require special consideration in developing the MSP.

In preparing the MSP, the following information should be gathered for the plan:

- port name, contact numbers;

- maps identifying the port location, access roads and building floor plans;

- number of craft dock points, and who else uses the facility (organizations/corporations/private individuals);

- interior layouts of the terminal indicating the location of security services, first aid, airline VIP areas, restrooms, telephones, restaurants, shops, foreign currency exchange stations, car service locations;

- coast guard alerts;

3. US Department of Transportation. *Port Security: A National Planning Guide*, US Coast Guard (1997).

- overall conditions of the port;

- size restrictions for crafts;

- operational hours and after hours requirements;

- past problems reported from prior visits;

- common waiting times for commercial transportation, luggage pickup, customs and car service.

Section 2
Response

INTRODUCTION

The EPS manager is responsible for locating, interviewing, selecting and training his/her staff. Unfortunately, the EPS manager is also responsible for terminating the employment of EPS team members when they fail to meet expectations, or it is felt by the principal that the member no longer meets with their approval. This section has two primary focuses, one on selection and the other on the training aspects that should be considered before selecting some-one or a plan developed to ensure that the EPS member receives the proper training. Figure 2 depicts the selection process and specific areas of training that should be accomplished.

SELECTION

Selection of EPS members is an area that is often difficult to accomplish due to the lack of qualified individuals. Managers are typically subjected to a barrage of resumes that usually exaggerate the applicant's qualifications. Although government sector agencies often possess the time and resources to select new members properly, the private sector typically does not. However, one private sector corporation decided to incorporate the same principles of selection in the development of their newly formed Executive Services Team. Meijer's Corporation,[4] a retail store chain in the USA, began using these 'formalized' principles to design the team's structure and

4. Meijer's Corporation, Results of Selection Process. Interview with Brian Kelly, Managing Director (2000).

FIGURE 2. Response

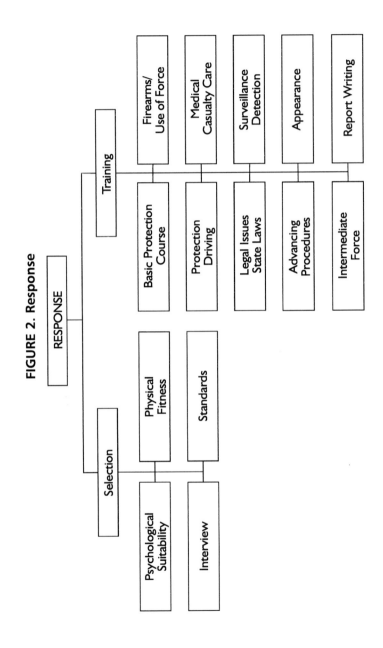

develop a criteria to select and train their members. As a primary consultant to Meijer's Corporation, the author elects to illustrate their process in this text.

Standards

Standards are the benchmarks that an individual is expected to achieve and maintain. Without standards the EPS has to fall back on his/her own knowledge of past training and experiences. The same holds true when deciding what standards are going to be followed during the selection process. Policies on selection should be designed to establish criteria for what is acceptable. Failure to follow specified standards sends a clear message to other EPS members that standards are there, but they are not important, or can be violated.

When writing standards the EPS manager should consider:

- the process by which a person will be selected – application/interview/testing;

- the desired qualifications for the position – trained/ untrained/military background/formal education (degree)/ work experience;

- the suitability of the individual to become a part of the existing team. May be a good idea to allow other team members to participate in the process;

- clear and understandable probation period and written expectations to ensure continuous progress.

Psychological Suitability

Conducting psychological testing and interviews can become expensive, but the results can discover potential problems before they become the EPS manager's problem. The work of the EPS is similar to that of a police officer.

It has been said that the job consists of long periods of boredom, broken up by a few moments of sheer terror. It is during these moments that the EPS manager wants to know that the applicant will respond according to the best options.

In considering suitability evaluation of this nature it is important to select a methodology, and also a professional who has experience in addressing the issues associated with the EPS position. Psychologists who conduct law enforcement evaluations would be a good place to start. They are typically keen when looking for signs of unsuitable behavior.

Physical Fitness

When selecting an EPS, his or her level of physical fitness must be observed. Total fitness is a combination of exercise, dietary control and mental attitude. An individual may look fit, but fitness evaluation can only be determined through a fitness test and medical examination. Again, the expense of a complete medical physical proves to be an up front investment, and the pay-off comes in understanding the person's ability to perform physically under stress. It is recommended that fitness evaluations be conducted prior to selecting the person for the team; if not accomplished before selection it may be difficult to justify terminating the employment of the person after selection, should he or she not qualify.

Interview

The interview process for most applicants is the same. The typical questions about likes and dislikes are asked, along with identifying strengths and weaknesses, but in selecting a specialist additional open ended questions should be asked.

Seasoned interviewers will have developed other techniques and questions to ask during the process. In addition to the questions and responses, it is imperative to observe how the applicant acts during the interview.

TRAINING

The training of an EPS is beyond doubt the single most important aspect of the profession. If we believe this to be true, then the question is, what type of training should an EPS have? Governmental EP teams[5] spend hundreds of hours in 'basic and advanced training' to ensure specific techniques are mastered through continuous practice. This is not necessarily the case in the private sector. A person who wants to join this profession, but has little or no formalized training seeks out opportunities to attend various training programs held throughout the world. Identifying the best training program that will accommodate the student as well as the potential client, yet train the student in the appropriate topics is not always easy. The next part of Section 2 is devoted to benchmarking the topics that should be taught in each of the course areas.

Basic Protection Course

After reviewing numerous protection courses and surveying professionals presently working in the public and private sectors, the following list of classes should be included in a basic protection program: history, interpersonal communications, basic movement skills and non-verbal communication.

Although these offer a good basis in the beginning, other recommended classes that could be in the program, depending on its length, include: protection driving, legal issues and state laws, advancing procedures, intermediate use of force, firearms/use of force, medical casualty care, surveillance detection, appearance and report writing.

Additional training topics could include: marine security, airport security, protocol, security surveys, threat assessment and specialized equipment training. It is incumbent upon the EP manager to ensure that the specialists receive updates on the latest techniques that provide for professional growth.

5. Governmental EP Team is defined as a team that focuses on the protection of a governmental official, includes US Cabinet Secretaries and Presidential teams.

Section 3
Organization

INTRODUCTION

In this section an overview of the type of issues concerning organization structure is outlined. A few years ago management consultants were advocating reducing the number of policies and procedures in the organization. In many cases, most policies weren't being followed or even known by employees. For a professional EP team, policies and procedures assist the specialist in making decisions in the manager's absence and provide the specialist with a comfort level of what is expected. Figure 3 depicts the elements that provide structure to an organization.

TEAM LEADERSHIP STRUCTURE

The EP manager must clearly delineate the structure of the organization beyond just the team itself. Individuals working close to the principal may possess some control over the team and, at times, provide direct supervision based on the venue or situation. Regardless of the structure, the manager should communicate it to the specialist and discuss any concerns the specialist may have. Policies can help provide direction to a specialist and executive staff, but a policy is only as effective as the principal allows it to be. As an example, any policy that is continuously and knowingly violated by the principal will not be taken seriously by the staff.

The principal should be consulted and informed about the policies to the greatest extent possible. The principal should be supportive of the policies developed by the EP manager.

FIGURE 3. Organization

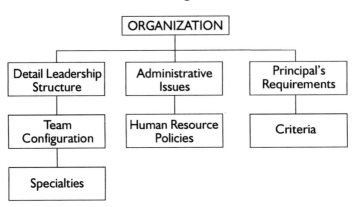

Team Configuration/Specialties

The EPS job description should identify the primary areas of responsibilities he/she is required to perform. As in any position, individuals on the team will probably possess different levels of expertise and be trained in a variety of specialty fields. The wise EP manager will use this information to the fullest extent. As an example, an EPS that has specialized training in firearms should be considered for the position of firearms instructor. Other specialties that are common to EPS who are well trained and experienced include:

- explosive/bomb recognition;

- evasive driving;

- surveillance detection;

- intelligence research;

- professional trained instructor;

- paramedic training.

ADMINISTRATIVE ISSUES

As in any organization, the EPS is a part of an overall organization and subject to the same policies as any other employee. The EP manager is responsible for dissemination and enforcement of administrative policies. As an example, the manager may be required to follow specific guidelines in promoting specialists, providing annual feedback through the performance appraisal process, and ensuring that other employee relation policies, like non-discrimination, are adhered to.

Human Resource Policies

The EPS can work for an organization in a variety of ways. First, the specialist can contract to the organization to render direct services. In this case the EP manager should possess a contract with the individual that outlines the rate of compensation, the specific time period services are to be provided, and any additional verbiage about non-disclosure agreements and associated liability. In these cases human resource policies don't usually apply. Second, the EPS can work directly for the organization or be an employee of a contract company contracted to provide EP services. In these cases human resource policies do apply. In the US companies that hire employees are required by law to provide certain information and benefits for their employees.

PRINCIPAL'S REQUIREMENTS

In addition to the human resource policy requirements, the principal will have his/her own personal set of requirements that should be established in policy format. The EP manager must develop a relationship with the principal that allows for open two-way communication. Being the manager of a team without this ability to discuss issues as they arise will be frustrating to the manager and team members. This does not mean that the manager needs to be in the 'boss's' office daily, but regularly scheduled briefings establish an opportunity for such interaction.

Criteria

The type of requirements set by the principal can include:

- set arrival and departure times for the office, travel (60 minutes before departure as an example) and events;

- preferred duress signal during escort;

- preference on close protection;

- protection issues related to family members;

- duty and responsibilities (ie, require EPS to attend certain outings with their children).

Section 4
Technical

INTRODUCTION

In this section the discussion focuses on the technical aspects related to managing the EP team (see Figure 4). The areas illustrated here are often contracted out to special service providers who have years of experience in their field. Although these services may be contracted, the EP manager should possess a basic understanding of the need for these technical aspects and when they should be applied. This is an area in which the EP manager can help his team grow by seeking out additional opportunities to receive training.

TECHNICAL SURVEILLANCE COUNTERMEASURES (TSCM)

TSCM or, as it is often referred to, 'debugging' can be an invaluable skill for an EP team to possess. Its purpose is to identify electronic surveillance devices used either to listen to, or visually observe the activities of another. Although the odds on having a device present are less than 2%, the time afforded to the effort is considerable. Devices of this nature are used to gather information or intelligence on an organization. They may not necessarily be difficult to obtain and install, but they are not the best means an adversary might use to collect information. More common ways include:

- searching through trash that has been disposed of;

- compromising support staff;

- computer intrusions;

- failure to identify sensitive documents and properly mark them;

- theft of laptops with data left on them.

FIGURE 4. Technical

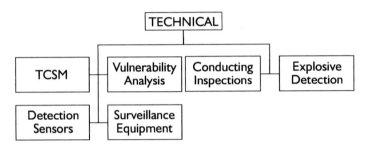

The EP manager should have available to him/her the contact information of a reputable TSCM specialist who can respond to such incidents. If TCSM services are required for long periods, consideration should be given to training and purchasing equipment for another team member. Since a decision to train someone to conduct a 'full survey' can be a sizable investment the following physical survey approach can be conducted without equipment:

1. Start the physical survey by identifying a starting point in the room and decide to work either clockwise or counterclockwise around it.

2. Move around the room critically examining each item to ensure that it is what it appears to be, that it operates as intended, that the weight is right. If required, x-ray the item.

3. Examine the walls looking for pinholes, patches over the wall, loose wallpaper or hollow sounding places in the wall.

4. Look at, touch and feel the surface of curtains and other fabric-covered objects. Squeeze the surface to ensure no devices have been installed.

5. Check door jams and associated hinges and other hardware to see if any scratches have occurred.

6. Examine the edge of the carpet or other flooring, including baseboards, to ensure no unusual wires or lumps exist under the flooring.

7. Examine each piece of furniture to see if any holes have been drilled out and freshly filled.

8. Open and remove each drawer and check under desks and cabinets.

9. Remove ceiling tiles and examine the area above the ceiling.

10. Remove all wall plugs, switches, pictures, windows and anything else and carefully examine them. If possible switch plate covers and other items in the room can be sealed with tamper indicating seals, with the date of the survey on the seal.

This type of physical non-technical surveillance, if conducted thoroughly, will identify the majority of devices, if there is one to be found.

VULNERABILITY ANALYSIS

The vulnerability analysis methodology or process is a systematic approach to identifying the organization's assets, the potential threats that may exist, and evaluating present security countermeasures. The EP manager will find that an understanding of the methodology will provide benefits beyond only the direct protection of the principal. By following the methodology illustrated below, the EP manager can identify the value of the assets, the potential threats, the effectiveness of existing security countermeasures and implement upgrades as required.

1. Characterize the facility to be analyzed. This involves a complete description of the organization's mission and relevant data, such as the location, number of employees and activities performed and physical building description.

2. Identify the assets possessed by the organization. Assets can include people, property and information.

3. Identify the type of threats that may exist. Security countermeasures cannot be installed for every possible situation; therefore, threats specific to the assets with a higher probability of occurrence must be identified.

4. Evaluate present level of physical protection systems in place. Understand that good security countermeasures include people, technology and policies all working together to produce an effective system.

5. Recommend upgrades to improve either part or all aspects of the security system. Some recommendations may include new policies, purchasing additional equipment, or training of security personnel.

DETECTION SENSORS

Detection is defined as the discovery of an adversary's action. This detection can occur by either overt or covert methods. In order to discover an adversary action, the following events must occur:

1. A sensor reacts to an abnormal occurrence and activates an alarm.

2. The information from the alarm and associated assessment sub-systems is reported.

3. A person assesses the alarm to determine its validity and, based on the assessment, responds to the alarm.

As straightforward as that may sound, in reality numerous problems can occur prior to the final response to an alarm.

Detection Sensors

The EP manager should be familiar with the basic principles associated with each of the following detection systems:

- passive sensors – detect some type of energy that is emitted by the adversary or detect the change of some natural field of energy caused by the adversary. As an example, a fence sensor that detects someone climbing it;

- active sensors – transmit some type of energy and detect a change in the received energy created by the presence of an adversary. Microwave sensors set around a perimeter are examples of an active sensor;

- covert sensors – are hidden from view; example, a buried ground sensor cable;

- visible sensors – are installed in plain view of the potential adversary. Although easier to repair, they may be susceptible to tampering;

- volumetric sensors – operate when an adversary enters the detection area. The area under detection is difficult for the adversary to detect; therefore attempts to circumvent the field are often difficult, if the sensor is installed properly.

Each type of sensor must be properly applied to meet the expectations of the EP manager.

Various sensor designs and applications are available from literally hundreds of vendors and manufacturers. The problems faced by the EP manager are which ones to

33

purchase for the correct application and whether they will meet expectations, either anticipated or unanticipated. One of the best means to determine whether the sensor will operate accordingly is to ask the vendor for a list of contacts that have used the equipment in a similar fashion. If he/she can't produce any contacts or is unwilling to assist, the caution flag should immediately be raised.

SURVEILLANCE EQUIPMENT

Surveillance technology, like the new sensor technology, has progressed over the past few years. The design of covert equipment is more advanced and is accessible to the general public. When contemplating purchasing surveillance equipment the same principles of identifying the needs or requirements exist. The EP manager will determine whether the installation should be covert or overt, recorded or non-recorded or watched continuously by an EPS.

In preparing to deploy surveillance equipment the EP manager should consider the following aspects of surveillance:

- Is there a need to use color or black and white images?

- Is the surveillance camera placed where it will see the adversary and cover the desired areas?

- Is the principal aware of the location of covert surveillance (this could be grounds for terminating the employment of the EP manager)?

- Is there enough lighting at various times of the day to produce the desired images?

- If recorded, how often and expensive will it be to change out the tapes (not necessary if digital, but more expensive initially)?

- What type of monitoring equipment will be used?

- Who will be assigned to monitor the image and how will they report unusual activity?

Although there are other points to consider, answering these questions will help identify the appropriate application of equipment.

CONDUCTING INSPECTIONS

The EP manager should be familiar with the different types of inspections that occur within the principal's business and residences. The following six-stage method will help identify problems and potential solutions:[6]

Stage 1 – identify the fire hazards;

Stage 2 – identify the people at risk;

Stage 3 – remove, reduce and replace fire hazards;

Stage 4 – assign a risk category (low/normal/high);

Stage 5 – determine whether the existing fire safety measures are adequate or require improvement;

Stage 6 – record, in writing, the findings of the assessment.

Other types of inspections the EP manager may become involved in include:

- life safety issues;

- hazardous material;

- sanitary health.

Checklists for each of these areas can be obtained from the local Health and Occupational Safety offices in the community.

6. Dailey, W. (2000) *A Guide to Fire Safety Management*. Perpetuity Press Ltd: United Kingdom.

Lastly, crime prevention through environment design often provides increased protection without the visible effects of security. As an example, large concrete planters in front of an office building appear attractive to the general public, but they serve as a substantial barrier to anyone attempting to drive a vehicle bomb close to the building.

EXPLOSIVE DETECTION

It has only been over the past few years that explosive detection technology has become commercially available for use by the public. Today these devices are still very expensive and cumbersome to use. Currently, the best, least expensive detection device available is the EPS. Although canines are better at detecting actual material, they must be handled with care and their ability to work is inhibited by rest periods. The EP manager should have established procedures that are to be followed during bomb threat searches, vehicle inspections and building inspections. Each procedure should identify the means by which searches/inspections are conducted, the type of items to look for, the actions to be taken if a suspect device or material is found, and the need to document in writing the circumstances surrounding the event.

Section 5
Emergency

INTRODUCTION

In addition to the emergency plans established by the EP manager, special consideration should be afforded to Command Post Operations, Workplace Issues and Evacuation. In this section each will be outlined and details on procedural information discussed see Figure 5).

COMMAND POST OPERATIONS

As the name implies, the Command Post (CP) is the center of all coordinating activities. The size of the CP and its location are dependent on the type of operation that is being coordinated. The CP operations can also be mobile, again depending on the type of operation being conducted. The CP is responsible for monitoring the principal's location and related communications. The CP coordinates shift activities and logistical support.

Communications

Probably the most important aspect of the CP is its ability to communicate effectively with the team in the field. The CP maintains movement logs that identify the time the principal moves, the location moved to and from and the route taken. The EP team is responsible for checking with the CP as outlined in the operational orders. If the team fails to check-in, the CP initiates the emergency notification checklist and dispatches a secondary EP team or law enforcement personnel.

FIGURE 5. Emergency

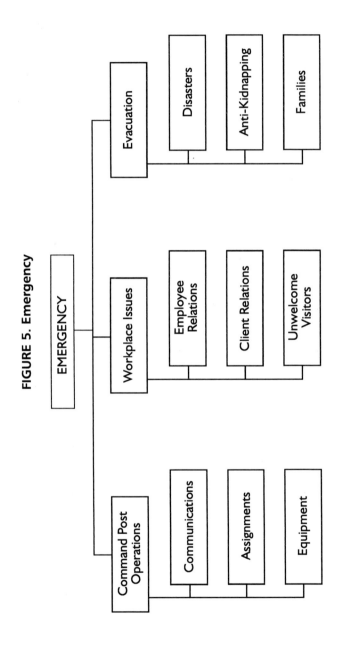

The EP manager develops a communication plan that lists the radio frequencies or channels to be used, the cellular phone numbers of all personnel, call signs used for the operation, and other pertinent information about communication, such as local emergency phone contacts and their numbers. If possible the sheet(s) can be reduced, reproduced and given to specialists in the field.

Assignments

The CP maintains a list of all personnel involved in the team operation. This includes pictures of the principal, his/her entour-age and specialists assigned. The CP is responsible for schedul-ing personnel and ensuring that they are posted at the proper location when needed. The CP will issue any special orders or checklists to specialists required to check identification of visitors. The CP assists the posted specialist in the event that an individual comes to the post desiring entrance but is not on the 'official' guest lists (mistakes have been known to happen). Daily, weekly and monthly itineraries are maintained at the CP and updated as necessary. This information should not be diss-eminated to all personnel. Personnel working in the CP should be specifically trained to fulfill the duty. Such personnel should be volunteers, rather than being forced to work the CP because of a disciplinary action or 'light duty'. The stress associated with the CP can be more intense than working in the team due to the various types of activities that are occurring simultaneously.

When specialists are not on an assignment or are on a break, it is important that the CP is not used as a lounge area for specialists to relax and 'kickback'. The CP should always be maintained in an orderly fashion and have the appearance of professionalism.

Equipment

For the CP to function in an efficient and effective manner the following list of equipment should be available:

- radios and cellular phones with spares;

- batteries (for flashlights/pagers/radios) and battery chargers;

- all operation plans for the team and associated route information and itineraries;

- CP logbook;

- computer with printer and extra printer cartridges;

- bulletin and erasable boards;

- office supplies;

- office furniture;

- spare keys to vehicles/residences/cabinets;

- weapons and ammunition stored in secure vault;

- first responder medical kits for the team;

- fire extinguishers;

- cameras and video equipment;

- smoke masks;

- tool kits;

- alarm detection and assessment equipment may also be stationed at CP.

Other items will be added to the CP as the need for specialized supplies and equipment arises. Examples include extra medication for the principal, handheld metal detectors and night vision devices.

WORKPLACE ISSUES

Concerns regarding violence in the workplace have grown over the past 10 years. The National Institute of Occupational Safety indicates that approximately 25% of all employees are threatened, discriminated against, or injured in the workplace annually due to violence. In 1998, over 1000 managers were killed in the workplace in the US. In this portion of Section 5, three areas of concern are illustrated for the EP manager's benefit.

Employee Relations

Employees constitute the majority of threats and attacks in the professional office workplace setting. The EP manager must have a clear understanding of the policies the principal's company follows in the event an employee is identified as a potential threat or is terminated under less than desirable conditions.

Other employee related policies the EP manager should be familiar with include:

- reporting of threats in the workplace by employees;

- how reports are disseminated to the EP manager;

- grievance procedures for employees and union rules on employee rights;

- termination procedures and related security measures, for example, taking the person's identification card;

- follow-up investigation process on incidents that occur.

41

Client Relations

Unfortunately, the principal's clients often express their dissatisfaction with the means in which the corporation treated them. This resentment is then expressed at the top of the executive ladder, the EP manager's principal. Threats can come in many forms: telephone calls, letters, e-mail and personal contact with the principal. One CEO of a major communications organization received threatening telephone calls about defective telephone equipment. When the caller felt nothing had been done to satisfy his complaint he mailed the defective telephone equipment to the CEO, again threatening him. The package was x-rayed and thought to be an explosive device, but it wasn't. Nevertheless, the EP team took the threats seriously and responded accordingly.

The EP manager is responsible for maintaining an organized filing system that records and tracks these types of threats. The type of information maintained includes:

- date and time communication was received;

- who was the first employee(s) to hear or see the communicated threat;

- what exactly was stated or written in the document;

- any indication of who may have sent the threat;

- any indication of a location sent from;

- identification of the means by which the threat would be accomplished.

Databases for cross-referencing this type of information are readily available and can be set up in a short period of time. A word of caution is appropriate, however, about maintaining this type of information: the database is only as good as the time devoted to it and the dedication of the person responsible for maintaining it.

Unwelcome Visitors

Unwelcome visitors can appear at the principal's office without warning. It is important that the EP manager determines the difference between the casual visitor and one who may have an actual interest in the principal. The visitor should be interviewed in a secure location. The manager should assume that he/she is potentially dangerous and precautions taken to ensure the safety of those in the area. The EP manager should note how the person is dressed and what personal information they are willing to offer, such as name and address and background data. The visitor should be identified, and surveillance video obtained and cross-referenced with known adversaries. The principal, his/her family and also members of staff should be shown pictures of the person in an effort to identify them or caution them to watch for this individual in the future.

The EP manager should allow the person time to talk and should listen carefully to their answers. Open-ended questioning will help gain a better understanding of the person's mental attitude.

In addition to verbal communication, non-verbal communication, such as hand gestures and facial expressions should be observed as well. Follow-up with local law enforcement agencies, institutions and the visitor's family may be warranted.

EVACUATION

The first step in determining whether or not to evacuate is to decide the credibility of the threat. The threat does not always include an adversary who wants to injure the principal. Some threats come in the form of weather and natural disasters, as in the case of the executive who was visiting a Caribbean island when a hurricane was reported to be closing in. In this case the EP team was not prepared for the type of reaction demonstrated by the island's other visitors and found themselves without resources.

Disasters

As previously identified, disasters pose a certain amount of danger that cannot be prepared for. Tornadoes, winter weather, hurricanes, strong winds, fire, etc., are often neglected in the evacuation planning effort. Since they occur infrequently the evacuation plan for these events is rarely exercised. The CP should, however, maintain constant contact with the national weather organizations and keep abreast of the latest developments that may affect the principal's movements. A place of shelter should be identified in the event that a disaster is immediate and comes unannounced. Transportation out of the affected area may be required. Everyone in the US has witnessed, at least on television, the massive number of people leaving the Florida Keys during a hurricane. Aircraft may not be available or even able to fly during bad weather so alternative means of transport should be established in advance.

Anti-Kidnapping

As discussed in the CMP, the EP team and principal should practice anti-kidnapping evacuation maneuvers. Acts of this type can occur at anytime, but the adversary(s) will normally act only when the probability of success is at its highest. Such circumstances include times when the EP team is at minimum staffing and coverage is not ideal. Evacuation plans for this type of event should include adversary attacks at the office, residence and during transportation. Each venue presents its own problems and mitigating responses.

Families

The principal's family should also be considered in the planning operation. Whenever possible, family members should exercise the various evacuation plans. Often family members are not under the control of the EP team. In these cases, the EP manager should develop a training plan that includes regular briefings to the family on security awareness and evacuation plans. The US Department of State has developed a complete security awareness program for family members preparing for an overseas assignment.

Section 6
Communications

INTRODUCTION

Communications can come in many forms: written and verbal, in person, over the telephone, by letter or e-mail. The number one problem faced by any organization is the lack of effective communication. In this section the means by which an EP team communicates will be discussed (see Figure 6). The files maintained by the EP team and how information is disseminated are important aspects of effective communication. Reports generated by the EP team should be reviewed by the manager and files created. Expert databases are available and are an excellent means to keep and disseminate information.

INCIDENT REPORTS

Incident reports are generated when an unusual event occurs that the specialist or the EP manager believes requires documentation. Specific types of events requiring reporting can be identified in a policy, although it is believed that any event the specialist feels needs to be recorded should be written in an incident report. It is best to develop a standard report form that best accommodates the EP team's needs. The basic information any report should contain includes:

- who;

- what;

- where;

- how;

- when;

- why.

FIGURE 6. Communications

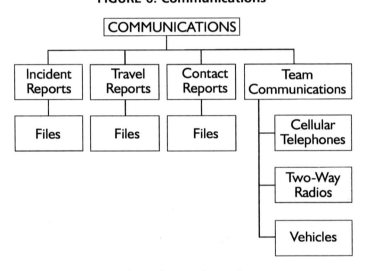

Reports are written in a clear and concise manner, addressing the facts as observed. Opinions are generally left out of incident reports, unless the writer's opinion is clearly stated as an opinion. Incident report summaries are helpful in identifying statistical analysis of the types of events that have occurred over a given period. As an example, the number of incident reports written on unwelcome visitors may have increased over the previous months.

In addition to tracking the number and type of incidents that occur, it is also possible to cross-reference specific data in the report with other available report information. Reports on advances conducted could be part of the cross-referenced information.

TRAVEL REPORTS

Travel reports provide details obtained from the various trips taken by the principal. This includes all advancing contacts, locations visited, people contacted and visited by the EP team and principal, event information and schedules. This information is invaluable to the EP manager as a means to prepare for future trips and provides excellent training material for newly assigned team members. Section 1 (under the Travel Plan heading) illustrates the type of information the EP manager can expect to see in the travel report.

CONTACT REPORTS

These are reports made by the EP team on individuals with whom the team or principal has made contact while traveling, or at work or social events. Normally recorded on 3 x 5 cards in the first instance, the information obtained can offer the principal critical information when required for a pre-briefing or impromptu occasion.

Contact reports should contain the following information:

- names of executives and others at the company facility;

- government personnel;

- persons the principal may have social engagements with at their home;

- helpful staff members, hotel staffers and drivers;

- special interests;

- a specific note on the contact, such as the wife has a long-term illness.

TEAM COMMUNICATIONS

The EP team uses a variety of instruments to maintain communications with each other and the CP. These types of communication are maintained expertly and the team receives regular training on how to use them. Failure to understand how they work and how they can be best deployed can prove fatal to the EP team during a crisis situation.

Cellular Telephones

Today's cellular phones and paging systems offer a wide range of accessories that assist the EP team in communicating. The features available on most cellular telephones include:

- contact number storage;

- automatic dialing capabilities;

- ear piece and lapel microphone;

- voice messaging, with remote paging;

- written messages in lieu of voice communication;

- single button duress capabilities.

Regardless of the type selected, it must be dependable, have an additional battery for backup and hands free operation when required. Remember, although listening to cellular telephone conversations is illegal, non-sophisticated devices can easily monitor them.

Two-way Radios

Two-way radios are often issued with little or no instruction. The EP manager may not realize that the team members may not be familiar with all of the radio's capabilities. These types of radios can be equipped with numerous channels or frequencies, private or secure voice capabilities, duress capabilities and accessories like an earphone and microphone. Call signs are often issued to the operator. Call sign determination may be assigned to specific radios for accountability before and after the issuing process.

Again, maintenance of the radio equipment is critical to the operation. Maintenance of the radio receiver and transmitter equipment is another critical area that should not be neglected.

Vehicles

The standard communication devices already discussed can be used while in a vehicle.[7] The setup for usage in a vehicle may be a little more complicated, but just as effective.

Vehicles can be equipped with additional communication devices that allow the vehicle to be tracked. Now available as an option in many luxury automobiles, these kinds of tracking devices are set up to monitor various activities. Activities that can be monitored include:

- vehicle location;

- fuel status;

- whether or not an authorized person is driving the vehicle;

- driver feedback with information on road construction, and other delay information;

- remote engine 'shut down' device;

- duress capabilities for the principal and driver.

7. Vehicle in this case is defined as any mode of transportation, ie, automobile, boat, plane, etc.

Section 7
Threat Assessment

INTRODUCTION

The basis for a EP team is the Threat Assessment (TA) methodology. The EP manager should, as a part of the regular duties performed, maintain information about the various types of threats that exist. The successful EP manager is one who has developed the ability to master the threat assessment methodology, has made the information collected easily retrievable, and can analyze the data effectively. In this section the threat assessment methodology will be discussed and the types of data collected outlined (see Figure 7).

INTELLIGENCE GATHERING

The gathering of data is not the difficult part of the intelligence field. Data is literally available at every turn, in newspapers, on television, through e-mail and of course from people. The problem that is quickly encountered is, firstly, how to manage the vast amounts of information, and then, secondly, how to interpret the data into valuable information to be used in the assessment methodology. The acronym D.R.I.P. illustrates what is meant: Data Rich Information Poor. In gathering data the best sources to begin with include:

- local crime analysis and related studies;

- professional organizations which maintain information about the principal's area of expertise or business;

- published literature.

FIGURE 7.Threat assessment

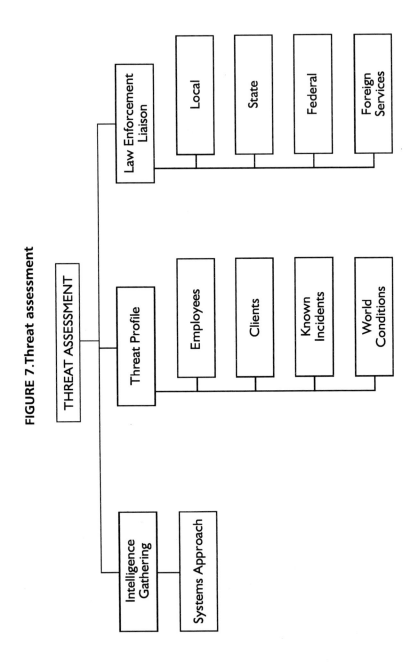

These sources can provide details about the current activities of individuals or groups, which may pose a threat to the principal or his/her related activities. The EP manager will want to develop a 'network' of human intelligence that can be called upon as needed, and may be more importantly, a network, who when they hear something of interest to the manager will call them to discuss the information they have collected.

Systems Approach To Intelligence

A systematic approach to gathering intelligence and maintaining the information should be created that accomplishes two primary objectives for the EP manager. The first objective is the collection and databasing of the information. Secondly, the information must be easily retrievable in a format that can be understood and which promotes dissemination to the EP team as needed. As an example, a US Department of Justice publication illustrates the type of information that may be considered in a threat assessment database.[8]

1. Identify the person(s) making the threat, this requires investigating the individual.

2. Identify the types of materials possessed by the individual, ie, magazines, letters, etc.

3. Identify persons who may know the person and have contact with them on a regular basis.

4. Obtain any previously recorded information about the person.

5. Lastly, identify the person's interest in the principal, the company represented by the principal, any type of

8. Fein, R.A., Vossekuil, B. and Holden, G.A. (1995) *Threat Assessment: An Approach To Prevent Targeted Violence*. National Institute of Justice: US Government Publication.

communicated threats, ownership of any weapons, and whether the person has been seen at any events attended by the principal.

When the person is unknown to the manager a different database is created to track the type of words used in the threat, the type of communication media used, date and time received and any other differentiating factors.

THREAT PROFILE

The term 'profile' has several different meanings, but for the purpose of this text the word profile will be used to illustrate the types of adversaries that may pose a danger to a principal, and the adversaries' motivations and capabilities. A 'design basis threat' (DBT) statement should be established that identifies the adversaries who pose a danger to the principal. By using a DBT statement the EP manager can plan the type of security needed for a given event or situation. The DBT is based on actual intelligence information gathered, which assists the EP manager in making protection level decisions.

In developing the DBT the following information is identified and considered:

1. Types of threats that have occurred in the past directed towards the principal or other persons in similar situations.

2. Any problems that have occurred at the workplace and who has been the target.

3. Review the client complaints and any negative reactions from them.

4. Any international relation concerns that the principal or organization may have.

5. Types of tactics used by adversaries and the means by which the threats were made.

6. The number of adversaries and the level of sophistication they require to accomplish their goals. As an example, were explosives used in their attack?

7. Types of equipment and information required by the adversary.

This information and more can be set in table format for clarity and easier dissemination. Specific threats identified during the analysis should be reported to local law enforcement for further official investigation.

LAW ENFORCEMENT LIAISON

The ability to network with the various law enforcement agencies is critical to gathering good information. Law enforcement agencies by nature are often restricted from releasing information, especially information that is required for an open investigation or upcoming court case. Law enforcement personnel are known to cooperate as much as legally possible and should not be asked to provide any unauthorized information.

Creating a network with local, county, state and federal agencies will help foster a good working relationship, without which the EP manager cannot be effective. Developing a similar network with foreign governments may not be as easy. Overseas law enforcement agencies often do not operate by the same rules. For some agencies bribes are often a way of business and they require payment before service. The decision an EP manager makes in these cases is to determine whether or not this is acceptable. The EP manager should not feel alone in making such decisions; multinational corporations struggle with this decision daily, trying to decide whether to pay for law enforcement response to a crime that affects their business. An example is a US based company which is having its trademark and products illegally reproduced in Asia. Local law enforcement will respond but only after some compensation from the company. Would you pay or not?

Section 8
Conclusion

Although much of the information written in this guide may
be obvious to the 'seasoned' professional EP manager, the
author's experience in dealing with a broad cross-section of
those responsible for this function indicates there is a need
for this information. It is anticipated that this guide will help
future managers understand the requirements for leading a
group of individuals in a noble profession. This guide
provides an introduction to the processes involved in
executive protection and is a good starting point for those
interested in the field.

Appendix A
Office Security Plan

The following strucural information should be considered:

- number of floors;

- number of tenants and types of activities they are involved in. It is important to identify the risks associated with external organizations that the principal could be exposed to;

- vacant areas;

- storerooms;

- construction underway or future planned construction;

- building owner's name, address, emergency contact information;

- points of contact at the building site, position, title, address and 24-hour contact information;

- occupant contact, position, title, address and 24-hour contact information.

Appendix B
Resident Security Plan

The following strucural information should be considered:

- number of floors;

- number of family members and guests that normally
  reside in the home and guesthouse(s). Identify the types of
  activities that they can be involved in while on the resident
  grounds, ie, exercise room, pool, riding stable, etc. It is
  important to identify the risks associated with activities the
  principal could be exposed to;

- remote areas on the grounds;

- equipment and storerooms;

- remodel or construction areas or future planned
  construction;

- residence manager's name, address, emergency contact
  information;

- contact numbers of all buildings on the grounds and
  people responsible for each building or activity function,
  to include 24-hour contact information.

Appendix C
High Threat Operations Plan

The following elements should be considered:

- contact procedures for the Command Post location and local law enforcement;

- pre-arranged safe haven locations (corporate locations with onsite security, law enforcement offices, etc.);

- checklist for contacting crisis management team regarding the increased threat;

- list of names of corporate personnel who are available to supplement EPS team. This often means enlisting non-security personnel to serve less critical duties. These individuals should have received at least minimal training in which duties they could be asked to perform.

Appendix D
Airport Security Plan

The type of information contained in an ASP incudes:

- airport name, contact numbers;

- maps identifying the specific layout and floor plan diagrams;

- number of terminals and airlines using the facility;

- interior layouts of the terminal indicating the location of security services, first aid, airline VIP areas, restrooms, telephones, restaurants, shops, foreign currency exchange stations, car service locations;

- Federal Aviation Administration alerts;

- overall conditions of the airport;

- landing and take-off restrictions;

- hours airport is open and closed;

- past problems reported from prior visits;

- common waiting times for taxing, luggage pickup, customs and car service.

In dealing with private and corporate aviation facilities additional information should be obtained. In addition to the aforementioned information it is important to identify:

- night-time operations allowed;

- restrictions on certain aircraft;

- length of the runway;

- mechanic available;

- fuel;

- security services – patrols/stationary posts;

- catering services;

- meeting facilities;

- secure parking for aircraft;

- emergency response crews at the facility (ambulance/fire);

- de-icing capabilities;

- snow removal equipment;

- clearance for plane side pickup;

- law enforcement jurisdiction and response capabilities (contact numbers).

Appendix E
Special Events Planning

The following should be considered:

- a minimum of three EPS should be used: the team leader is with the principal; the second EPS is used to coordinate equipment requirements and provide relief for shift work; the third specialist conducts advance work;

- the EPS manager should never place him/herself in charge of all the event's security needs;

- obtain a complete itinerary of the event which should detail most of the required information to be used in the SEP;

- event credentials are designed to be unique, printed on both sides and laminated, to reduce reproducing the access badge. Those with direct access to the principal should be given special access codes, such as a gold star pin;

- other identification devices to consider include wristbands, additional laminates or pins.

When meeting with the event organizers the EPS manager will want to know the overall security plan that has been developed well in advance so that any areas of concern can be addressed prior to the event.

General information gathered for the SEP includes:

- the type of event to be held;

- the anticipated size of the event, and the type of people anticipated to attend (for example, a rock concert or a night at the opera);

- location of 'Safe Rooms' and first aid stations;

- ability to acquire something to eat and drink, and menu selection, if the principal has any special dietary needs;

- safety aspects associated with the stage, podium and fire evacuations;

- arrival/departure route security to ensure that fans or potential adversaries aren't waiting to attempt contact with the principal.

Appendix F
Security Route Plan

The following information should be included:

- location/date/time/city and county, if overseas include country;

- departure/arrival points and anticipated time movements occur;

- the type of resources required for movements – number of vehicles/EPS members;

- location of 'Safe Places' such as fire stations/police stations/hospitals;

- alternative routes – highways/residential/county roads;

- condition of streets traveled – construction/paved/dirt roads;

- identify traffic flows – light/moderate/heavy, the time of day, and whether it is a holiday or weekend;

- identify danger zones and choke points – areas at which traffic slows/stops and funnels vehicle movement into one specific route without alternatives;

- locate overpasses – people have been known to throw things off them;

- railroad crossings and tunnels;

- high ground/culverts/pedestrian crossings;

- businesses and buildings along the route – principals have been known to want to stop at a store;

- schools;

- identify any special events that may be occurring or planned in the future;

- assign route checkpoints and assign code designators;

- identify dead zones for radio communications.

Acknowledgements

The author sends a special word of thanks to Jerry Glazebrook for his advice, assistance and friendship in writing this guide.

The publishers would like to thank Tenzi Luca, Giovanni Manunta and Anthony Ricci.

Commissioning Editor: Dr Martin Gill, Scarman Centre, University of Leicester, 154 Upper New Walk, Leicester, LE1 7QA, UK (Telephone: +44 (0) 116 252 5709; Fax: +44 (0) 116 252 5766; e-mail: mg26@le.ac.uk).